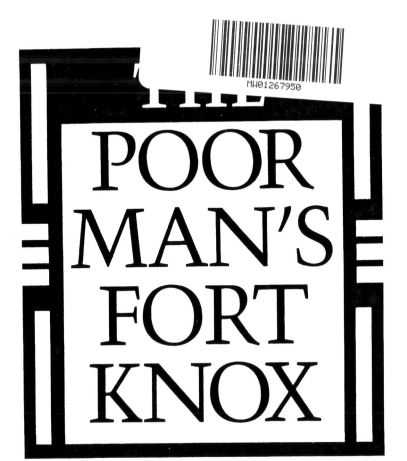

WARNING

Technical data presented here on safes, strongboxes, and related devices, as well as the law, legal information, and methods of protecting personal property inevitably reflect the author's individual beliefs and experience with specific equipment under specific circumstances that the reader cannot duplicate exactly. Therefore, the information in this book should be used for guidance only and approached with great caution. Legal advice should be obtained from a licensed attorney. Neither the author nor the publisher assumes any responsibility for the use or misuse of information contained in this book. This book is *for information purposes only.*

THE POOR MAN'S FORT KNOX

Home Security with Inexpensive Safes

Duncan Long

PALADIN PRESS
BOULDER, COLORADO

Also by Duncan Long:

AK47: The Complete Kalashnikov Family of Assault Rifles
AR-15/M16 Super Systems
The AR-15/M16: A Practical Guide
Assault Pistols, Rifles, and Submachine Guns
Build Your Own AR-15
Combat Revolvers
The Complete AR-15/M16 Sourcebook: Revised and Updated Edition
FN-FAL Rifle
Hand Cannons: The World's Most Powerful Handguns
Homemade Ammo
Mini-14 Super Systems
The Mini-14: The Plinker, Hunter, Assault, and Everything Else Rifle
Ruger .22 Automatic Pistol
SKS Type Carbines
Streetsweepers: The Complete Book of Combat Shotguns
The Sturm, Ruger 10/22 Rifle and .44 Magnum Carbine
Super Shotguns
The Terrifying Three: Uzi, Ingram, and Intratec Weapons Families

The Poor Man's Fort Knox: Home Security with Inexpensive Safes
by Duncan Long

Copyright © 1991 by Duncan Long

ISBN 0-87364-645-2
Printed in the United States of America

Published by Paladin Press, a division of
Paladin Enterprises, Inc.
Gunbarrel Tech Center
7077 Winchester Circle
Boulder, Colorado 80301 USA
+1.303.443.7250

Direct inquiries and/or orders to the above address.

PALADIN, PALADIN PRESS, and the "horse head" design
are trademarks belonging to Paladin Enterprises and
registered in United States Patent and Trademark Office.

All rights reserved. Except for use in a review, no
portion of this book may be reproduced in any form
without the express written permission of the publisher.

Neither the author nor the publisher assumes
any responsibility for the use or misuse of
information contained in this book.

Visit our Web site at www.paladin-press.com

CONTENTS

Introduction
1

Increasing Your Safety with a Refuge Room
5

The Low End
15

The High End
29

Conclusion
41

ACKNOWLEDGMENTS

Thanks must go to the companies listed in this book for supplying me with information about their safes and other devices for protecting the homeowner from thieves, vandals, and government snoops.

The people who helped produce this book are much appreciated, too: Jon Ford, who "walked" the idea through its paces, and Peder Lund, who gave the go-ahead for another of the many fine titles Paladin Press offers. Thanks must also go to editors Karen Pochert and Donna DuVall, who slaved over the manuscript to ferret out mistakes and make it more readable, and all the others on the Paladin staff who continue to do their marvelous magic of turning an author's rough work into printed form.

And the usual tip of the hat to Maggie, Kristen, and Nicholas.

INTRODUCTION

Why would anyone need a home safe?

Why wouldn't they! Anyone living in a high-crime area can tell you why. Having anything of value that can be easily carted away by thieves means you *won't* have it for long in many areas of the world, including the United States. In fact, according to Jeff Cooper in his "Cooper's Corner" column in the April 1991 issue of *Guns & Ammo*, the police in Drew Price (a suburb of Los Angeles) recently recommended that homeowners "not keep their valuables in their homes." This may be a solution for some—but not for most of us. This isn't a problem peculiar to California or even the United States, either; crime is on the rise almost everywhere.

Banks?

They aren't the perfect solution. The current rash of bank and savings and loans closures often leaves those with their money in savings or checking accounts high and dry for days or weeks. And those with their "valuables" in safe-deposit boxes are out of luck for even longer periods of time if a savings institution fails. Having cash, jewelry, or valuable papers at home in a safe can prevent becoming a pauper overnight (even if only for a few days) until the Federal Deposit Insurance Corporation (FDIC) gets around to reopening your bank.

Even if your bank is rock solid, it closes at 5:00 P.M. and doesn't open until the next morning or following Monday— not too handy if you need to access your safe-deposit box a couple of hours after closing time. There are also some things you can't very easily keep in a safe-deposit box—like firearms, rare coins, or other collectibles that often are *not* covered by the bank's insurance. And don't think safe-deposit boxes are never raided by criminals; while such occurrences are rare, they do happen. The author has an uncle who lost his coin collection in just that manner.

In addition to sometimes failing to protect their citizens adequately from crime, governments also often treat the law-abiding like criminals. The Internal Revenue Service goes through bank safe-deposit boxes from time to time. This can create problems even for those who are obeying the law, since cash or other valuables are often assumed to be locked out of sight to avoid paying taxes on them. More than one taxpayer has had to sweat it out when the IRS discovered his perfectly innocent nest egg and assumed he must be running an illegal business on the side. A home safe can prevent this situation.

Furthermore, local, state, and federal governments often pass laws that transform law-abiding citizens into criminals overnight. Legislation covering drugs, firearms, or you-name-it can suddenly make the items illegal, leaving owners of the banned materials in a moral quagmire. For such people, having a safe can be a temporary solution to the problem by buying some time, at least until it appears that the unfair laws won't be repealed.

Additionally, a safe can protect you from private and government snoops who seek to invade your privacy. More than one house has been surreptitiously broken into without a search warrant. Often, if questionable material is found, the information about it is exchanged, sold, or bartered to the police and a search warrant is obtained under other pretenses (like the proverbial "anonymous tipster"). Having questionable materials such as how-to manuals on machine gun conversions, drug manufacture, or other controversial subjects, or objects like dummy machine guns, grenades, or rockets, might subject the owner to a long legal hassle. Therefore, many who legally own these publications and replicas hide them in safes simply to avoid such problems, even though they've done nothing illegal.

In any or all of these cases, having a home safe makes sense. Now let's take a look at the least expensive ways to protect what you own and secure your privacy.

INCREASING YOUR SAFETY WITH A REFUGE ROOM

Most home safes cost an arm and a leg and are out of the reach of many pocketbooks. Even the most expensive safes can be "cracked" by a criminal with enough time on his hands or a police officer with a search warrant. But if you do purchase one, there are strategies that you can employ to minimize these problems and increase the protection offered by a safe.

In regard to the cost of the safe, you need to sit down and make a few mental guesses and calculations. How often do you expect to have someone break into your home and steal your valuables? Would they be able to work for a lengthy time, only a few hours, or just minutes before neighbors called the police or you came home from work? And how skilled would the criminals likely be? Would they be pros after expensive jewels, teenagers after a few bucks, or junkies after your color television?

Unless you're facing a seasoned pro after expensive jewels, your prized gun collection, or large amounts of cash (or contraband), you probably don't need the most expensive safes. A superexpensive home safe may make sense if you're a jeweler or keep sensitive company or government documents in your home, but for most people this isn't the case. Consequently, most readers would probably be better off spending money on homeowner's insurance (or increasing the coverage of a current policy) and purchasing a less-expensive safe for the few firearms, family heirlooms, or whatever they want to provide with extra protection.

Such insurance will protect big-ticket items you can't put in the safe, like your stereo, TV, VCR, and so forth. It makes little sense to purchase an expensive safe unless you have something considerably more costly to lock in it that would be excessively hard to insure. (If you go the "insurance route," be sure to videotape or photograph everything you have of value. Leave the tape or photographs in a sealed envelope with a relative or in some other safe place. Then, if you have a break-in, fire, or other disaster, you'll have a handy record to show to your insurance adjuster.)

If you're foolish enough to depart on an extended vacation and leave the phone on the hook, have newspapers pile up in the driveway, or announce your trip in the local paper, criminals will raid your home at their leisure. Expect to lose almost everything in such an event, since the burglars will know they have plenty of time to loot your home.

On the other hand, if they think you might be back momentarily, thieves will be working "under the clock" after breaking into your home. They'll miss things that are hidden, and they won't have time to fool with breaking into even a cheap safe. That means anything you can do to make their task harder will reduce your losses and might even make them decide it isn't worth the chance of getting caught to break into your home in the first place.

One way to do this is to make it appear as if you're at home even when you're not. If they still break into your home, they'll waste time making sure you're not in the bathroom before actually entering the premises. Once in, they'll have to work as fast as possible since it will appear you might be right back.

How can you work this head game on would-be burglars?

There are a number of ways to create this "insecurity complex" in the mind of a break-in artist. Among the best tactics when you're out of the house are to leave the phone off the hook, keep the radio or television turned on (louder than usual if possible), and have a few lights shining in rooms easily viewed from the street. These will all give a criminal pause. His question will be: is the owner at home or just not answering the door?

Of course no one wants to be shot, so add to a criminal's mental trauma by showing that you're armed and ready to take action. To do just that, an inexpensive sticker that warns criminals they may be shot can be placed on the front and back doors. The sticker need not be elaborate or overly specific as to an intruder's fate (such as the famous, "Trespassers Will Be Shot . . . Survivors Will Be Shot Again")—the National Rifle Association has a nice membership decal that doesn't make any actual threats (helpful if you ever have to defend yourself in court for having aced a felon in your house) but lets the criminal know that you're "armed and dangerous." The deterrence offered by these decals is worth the price of membership to the NRA, even if you aren't interested in guns or hunting. (As of this writing, a $25 membership to the NRA entitles you to, among other benefits, a year's subscription to *American Rifleman* or *American Hunter* magazine along with the intimidating decals. Write to the following address for information: NRA, Membership Division, 1600 Rhode Island Ave. NW, Washington, DC 20036.)

Since it takes time to "case a joint" to see if anyone is present before breaking into a house, there are a few things you can do to make this task as risky as possible for a burglar. One is to install automatic lights with integral photoelectric cells and motion detectors on your front and back entrances. Sears, Radio Shack, and other stores offer these units for $20 to $40, and they easily replace existing outdoor lights.

It is also prudent to trim shrubs so they don't afford hiding spots along the house, and having prickly plants like cactus or roses under windows can turn the area into the equivalent of a barbed-wire barrier (just don't forget to leave a clear space somewhere in case you need to exit during a fire).

Quality locks are important on your home and can pay big dividends by slowing a criminal's progress. A burglar knows that the longer he spends outside and the more noise he makes entering, the more apt he is to attract attention. Consequently, he may not bother to break in if the risks seem too great. Therefore, adding dead-bolt locks is a sound investment for protecting the valuables inside your home. Cross bars or other barriers that can be utilized when you are home give you added security.

Since burglars often enter through windows, a little thought should be given to these as well. Generally, the more layers of glass you have in a window, the tougher a job the intruder will have entering. Adding a bead of caulking around the top and side edges of storm windows makes them almost impossible to remove quietly, and double-paned windows can help in keeping more than the cold out. Drilling a hole in the window frame where the upper and lower halves meet will enable you to place a bolt through the frame to lock the window shut. (Don't be tempted to nail windows shut or install iron bars over them in a bedroom— you're apt to become a crispy critter if you have a fire.)

Living in an area with a low crime rate is an important consideration to minimize potential losses. If you're moving, take some time to ascertain which areas of a city suffer the most from break-ins. Sometimes you can drive a little further to work and live in an almost crime-free area. (The author knows of several communities where crime is so low that many people don't even bother to lock their homes when they go shopping during the day.)

Of course some criminals may ignore your decal, brave

the thorns, ignore the lights, and come right on into your home even if you're there. This is especially true in cities and states where gun control has disarmed many citizens or where criminals don't seem to mind killing or being injured. In such cases, it makes sense to have what has become known as a "refuge room" where you, your family, and your safe can enjoy an extra line of defense before a criminal can reach you. While refuge rooms can be quite elaborate and expensive, a little inexpensive do-it-yourself work can make a location capable of providing major protection to you and your valuables.

The first order of business in creating a refuge room is to place it where you can reach it in a hurry. This will allow you to enjoy the protection it offers *and* protect the room and all in it if you need to. Since many break-ins occur at night, it's wise to locate your refuge room close to your sleeping quarters or even turn your bedroom into such a "fortress."

Obviously your refuge room should have a door that can be locked quickly from the inside. Ideally the room will also have a phone on a separate line so someone can't disable it by lifting any extension phones off the hook after breaking in. Since many thieves also cut phone lines before entering, it's also wise to "armor" the line outside by having an electrician encase it in conduit (do-it-yourselfers can easily handle this work for only a few dollars).

An extra amount of safety can be realized by having a memory phone that can dial 911 or other emergency numbers with the touch of a button. Even though the police can be slow in arriving, this can give those willing to defend themselves a legal edge since the police will record the action going on in your room. Just be sure to voice your concerns that the person trying to get into your room is probably going to kill you (don't get too hammy), and be sure *not* to say anything incriminating like, "Die you stupid bastard" or "I'm going to put you out of your misery!"

Ideally your phone call should tell the police your address, where you are in the house, and what's happening, although this isn't essential in many areas since most police departments can trace a call in a matter of seconds. If shooting occurs or you have a gun, it's also wise to describe yourself—homeowners have been killed when mistaken for felons. It's a good idea to lay down your firearm *before* meeting the police at the front door.

While most thieves will head for the hills when the police arrive, they do get trapped at the scene of the crime occasionally. Therefore, have a spare key to the front door available in the safe room; it should be attached to a super large key chain so you can toss it to the police from inside your room. Let the law enforcement officers open the door and search the house—they're trained to do it right.

Since the principal purpose of the refuge room is to keep those who have broken into your house away from you for as long as possible, you need a massive, solid wood door with a heavy bolt and/or lock with its hinges on the *inside* so they can't be attacked from outside. Since most doors can be kicked in with repeated efforts, you need to reinforce the door from about waist level down with a metal plate or purchase a metal door designed for exterior use.

While the door to your refuge room can be one of those designed for walk-in safes (more on these later), most of us can't afford that much expense or prefer to avoid the odd sight a bedroom suite encased in a safe-style vault presents to visitors. Therefore you should consider keeping a firearm in your refuge room since a determined criminal will have gained entry to it before the police arrive.

In many areas of the country, it is possible to warn an intruder and—if he has you cornered and appears to have bodily harm in mind—you can fire through the door of the bedroom to protect yourself. But these laws vary considerably from place to place, so before you're in a

Increasing Your Safety with a Refuge Room

situation where you need to fire shots, contact your state or county attorney (who can tell you the law without charging you a legal fee). It's better to know the law *before* you need to shoot.

A refuge room needs more than a strong door. Most indoor walls are constructed of two layers of Sheetrock nailed to two-by-four wall studs spaced a foot or more apart. Any criminal who hasn't smoked his brain with dope (not always the case) will know that simply kicking at the areas between studs will enable him to produce a hole he can walk through to gain entrance in short order. Therefore, you'll need to "armor" at least the wall around the door and, ideally, anywhere that someone might kick through to enter from an adjoining room.

The easiest way to do this is to have the room paneled with a stronger material such as plywood, thick fiberboard, or similar building materials. The task is easy for those willing to read up on it (local libraries are a wealth of such how-to material). Standard paneling can also serve this purpose and can give a finished look with a minimum of work (though it's a bit more expensive). When using any of these materials, remember to secure them with considerably more nails than is normally called for on such jobs. For added strength, paneling can be placed over Sheetrock and/or a plywood subpaneling.

For a maximum-strength refuge room, concrete is ideal since it provides not only protection from kicking in but also affords varying degrees of security from bullets the intruder may try to send into your fortress. Of course cement and iron reinforcement rods are expensive and, in many houses, the weight can't be supported by the structure of the building. But it is something to keep in mind when house shopping or if you have a generous budget to work with. (Cement cinder blocks coupled with fiberglass/cement surface bonding materials like Quickwall might put this into the financial range of some do-it-yourselfers, but it would be a major undertaking best attacked only by those who know what they're doing.)

A higher level of security can also be afforded by an inexpensive burglar alarm. While units that attach to all the doors and windows are less apt to give false alarms, motion detectors or infrared alarms are considerably cheaper and easier to install. You can usually pick one of these up for a small price at your local Radio Shack outlet.

Once in place, an alarm can give you a few extra seconds to realize someone is in the house. For those with families, an alarm is nearly essential so everyone can be rounded up and secured in the refuge room before the door is locked. In such a situation, even more time can be bought by adding a hall door between the sleeping quarters and the rest of the house, then mounting an alarm so it will be set off before the hall door is penetrated. By the time the thieves get through the hall door, the alarm should have warned everyone to scurry into the refuge room.

Since you may not be home when a burglar breaks in, your refuge room should be capable of being locked from the outside. If you keep your safe within, this will give it an extra level of security since the burglar will waste time getting through the door and will have less time to spend trying to crack the safe. So the money you spend creating a refuge room can be a very good investment, especially given the cost of the higher-quality safes.

Of course, you can also protect your valuables by hiding them. While this isn't nearly as convenient as a home safe, it does have some pluses in terms of money. Although any thief with enough time will eventually discover any hiding place you might create, as noted above, such time may not be available to a criminal.

You might even employ all the above tactics *and* hide your refuge room. While hiding a whole room isn't easy, it is possible. For a more detailed look at how to hide anything from a few dollar bills to whole rooms, check into

Jack Luger's *The Big Book of Secret Hiding Places* (available from Paladin Press, P.O. Box 1307, Boulder, CO 80306), Charles Robinson's *Construction of Secret Hiding Places* (Loompanics Unlimited, P.O. Box 1197, Port Townsend, WA 98368), Michael Conner's *How to Hide Anything* (Paladin Press), and David Krotz's *How to Hide Almost Anything* (Loompanics).

Now that we've taken some time to check out the strategies and considered how much protection you might need, let's take a look at the types of safes that are available to you.

THE LOW END

Inexpensive safes and gun chests secured with locks rather than numeric pads or combination dials offer an attractive alternative to expensive safes. While they don't afford the level of protection from fire or thieves that the more expensive safes do, they will thwart many amateur burglars and will even slow down or stop Cro-Magnon tactics if they are anchored in position or placed in a spot where they're hard to attack.

In addition to the plus of their low price, these safes are easy to transport and place due to their light weight. They can be shipped by freight inexpensively or purchased through many discount stores and brought home in a car or pickup truck. One of the smaller units offered by Homak can even be shipped via UPS right to your doorstep for just slightly over $100.

The secret to using these safes (perhaps more accurately called "strongboxes") is to anchor them securely to the wall with the long screws provided; this keeps them from being carted off or dragged into an open area to be worked over with hammers and pry bars. A good tactic is to also place these safes in a location where it will be awkward for a criminal to attack them with heavy break-in tools; the narrow end of a closet, for example, is ideal. More security can be created by framing the safe with two-by-fours or even embedding the unit in concrete, leaving only the door exposed.

The Poor Man's Fort Knox

Since the security chest may still be ripped from the wall (or the wall torn away and the studs cut to free the safe), the more weight that can be added to the cabinet, the harder it will be to move. Therefore, it's wise to store large quantities of ammunition in the bottom of these units or, if you don't shoot, simply pick up some old wheel weights from an automotive store. These weights are often free (since lead isn't added to gasoline or paint these days), and a bucket of them will create some serious weight that will guarantee a hernia operation for your burglar if he tries to move the strongbox far.

Most of these security cabinets are designed for guns, but you certainly don't need to limit yourself to putting firearms in them. Units without internal barrel rests and the like have a nice open interior in which to stack things. Building plywood shelves inside the unit will help maximize the space you have to work with, and the addition of hooks will enable you to hang cameras, jewelry, or the like inside if you need to.

Foam padding (available free from many businesses selling equipment that is shipped to stores, such as computers or stereo equipment) strategically glued inside the cabinet will help keep your treasures from getting scuffed up. Scraps of carpeting will also work fairly well for this purpose and are generally available at construction sites or carpet stores for little or nothing.

The most competitively priced home security cabinets are offered by Homak Manufacturing Company (3800 W. 45th St., Chicago, IL 60632). The dimensions of the Homak security boxes and cabinets vary from pistol size (Model 3085, costing under $40 and designed for police stations as well as for the homeowner) up to a massive cabinet (the 3090, costing around $250) designed to hold sixteen rifles or shotguns along with a number of pistols and a large quantity of ammunition. Since each of these units is designed to be screwed into wall studs, and since several

The Low End

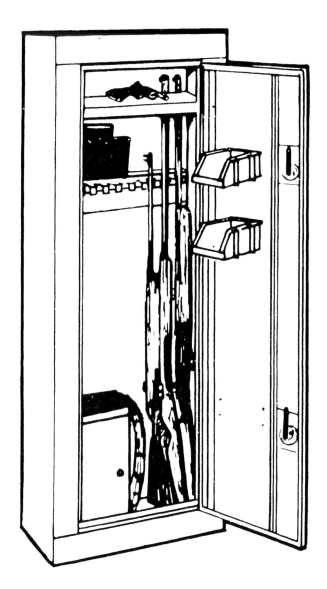

The Homak Model 3000, like many of the other security cabinets in the company's line, features 16-gauge steel walls and double antidrill locks. It affords a lot of protection for very little cash. (Illustration courtesy of Homak Manufacturing Company, Inc.)

can be connected together, you'll get more security if you buy several smaller units rather than one massive security chest. This will force thieves to open each one—a time-consuming proposition requiring a lot of work.

Homak's pistol-sized strongboxes have an added plus for those facing a court-approved search warrant. While the law is vague, most warrants cover only what is listed on the search warrant. Thus, if the search were being conducted for an illegal "assault rifle," for example, the pistol boxes should not be opened since they couldn't contain the large weapon. While the police often have a few tricks up their sleeves to circumvent the rights of the person under a search warrant (such as saying they thought they smelled the odor of dope emanating from one of the pistol boxes), a pistol safe does afford a little extra security during a warranted search. (It should be noted, however, that any safe can be ordered opened if a search warrant has been issued and the object being hunted for could be contained in the space in the safe. Therefore, more security from legal searches can be enjoyed with the expedient measure of hiding the safe itself so it can't be found. You can't be ordered to open what no one knows exists.)

The Homak Model 3040 security chest can be shipped via UPS. Its interior is small—rated for only four guns—but a lot can be stacked into the unit, which can be purchased complete with shipping for under $150.

For those wanting a safe in their pickup truck, Homak has even created the Model 3075 designed to fit behind a truck seat. Though not as strong as their other safes due to its single lock, it offers considerably more protection for firearms or other valuables than a glove compartment or even a locked trunk. This keyed unit might also be utilized inside a large car trunk, under the back seat of a car, or in a recreational vehicle.

All Homak strongboxes are constructed of 16-gauge steel that's welded together with a long hinge down their

The Low End

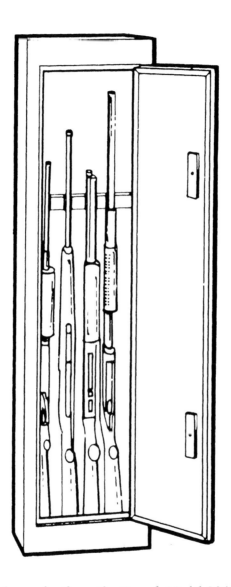

Like its larger brothers, the Homak Model 3040 should be anchored to wall studs to keep it put; the security cabinet comes with the hardware to do this. (Illustration courtesy of Homak Manufacturing Company, Inc.)

The Poor Man's Fort Knox

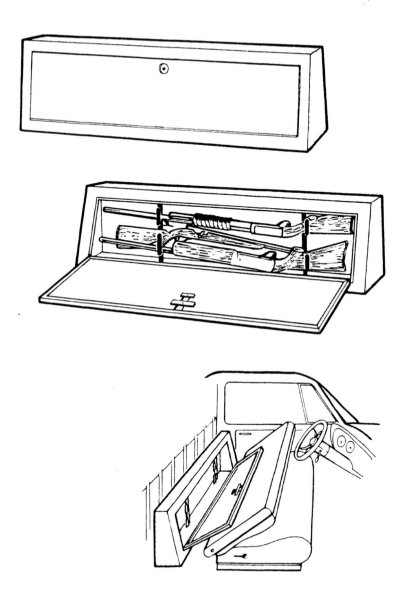

The Homak Model 3075 gives the purchaser a mobile security cabinet for his pickup or other vehicle. (Illustration courtesy of Homak Manufacturing Company, Inc.)

recessed doors. They have hardened antidrill locks, with a single lock on the smaller boxes and dual locks, keyed identically, on the larger units. If you order several of these security chests directly from Homak, it's possible to have them keyed all the same. Provided you're careful not to lose the key or leave it where a burglar can find it, this will make it easy to open the chests and still give you a large degree of protection.

At the high end of Homak's line is the Model 3990CL security cabinet. Unlike its predecessors, this unit features a combination lock rather than twin-keyed locks, making it approach a safe in its abilities while still carrying a relatively low price tag of $559. With its 14.7 cubic foot interior, it can store a lot of "stuff." Those unable to afford a safe who want more protection than is offered by standard security cabinets should consider the 3990CL.

One thing Homak safes don't offer is protection from fire. However, many business supply stores and even some discount stores offer a box that will protect papers during a fire. One of these relatively inexpensive boxes can be placed inside a strongbox to give you a low-cost alternative to an expensive fireproof safe.

If all you want is a fireproof place for papers, there's an even cheaper alternative, first suggested to the author by an insurance adjuster who noted that one part of a house is almost never damaged in a fire. The most fire-resistant area in most homes is the inside of the freezer since it's so highly insulated. Thus if you wish to protect papers, the least expensive solution is to purchase a Tupperware container that will hold your papers and "file" them in the back of the old refrigerator.

Oddly enough, while a brand new quality safe costs thousands of dollars, its resale value can be almost nonexistent. Therefore, if you live in a large city, you can sometimes locate an old safe at an auction house or estate sale that will only cost a few hundred dollars or less. These are

The Poor Man's Fort Knox

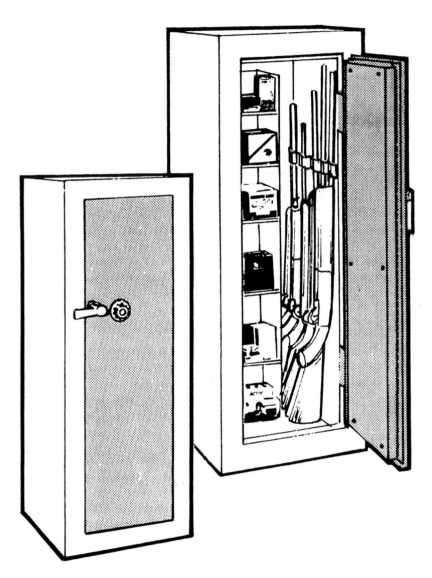

Homak's Model 3990CL security cabinet features a combination lock and can fulfill many duties handled by safes costing four times as much as its $559 price tag. (Illustration courtesy of Homak Manufacturing Company, Inc.)

often large, very burglar resistant, and extremely fireproof. The catch comes in moving the thousand-pound-plus safe to your home. But generally it's worth the effort if you wish to have a truly secure safe at a bargain-basement price.

Some of these old safes are sold "as is" with the combination to the safe unknown (you'll likely want to change the combination if it is known anyway). In such a case you'll need to have a *skilled* locksmith work on the safe, probably drilling it and replacing the combination lock if he doesn't have a recalibration key for it. Best bet is to contact the manufacturer of the safe and ask whom they recommend for the work. If the manufacturer is out of business, check with the local police department.

Once you have the safe in place, it should be surrounded in concrete or other materials so it is impossible to move. While these things are heavy, the lure of the wealth they may contain often causes groups of criminals to expend a lot of effort in lugging one off.

If you really want to "go on the cheap," one of the most secure places you can hide valuables is still in the ground. Most fires can't damage valuables in an underground cache, and criminals aren't likely to spend their day digging out in the open on the off chance something of value is there. The catch comes in the fact that it's very inconvenient to retrieve things buried in the backyard—and if you die, your heirs may never locate your treasure!

For those who wish to try this ultra low-cost method of hiding things, all you need is a shovel, a spot out of your neighbors' sight, and a watertight container to protect your valuables from moisture. If you're hiding a gun or some other item that is sensitive to rust or moisture, you can't be too careful—several chemicals formulated for protecting firearms against rust will also be useful. Outers' Metal Seal is one that's readily available at most gun stores; the chemical displaces water somewhat, adheres to the surface of steel, and acts as a lubricant as well.

For really long-term storage, metal parts can be encased in paraffin, Cosmoline, or some other grease and/or wax combination to prevent rust formation. This protectant can be improvised by melting candles and mixing it with a heavy oil, taking care not to set the mixture aflame. An easier alternative is simply to buy Outers' Gun Grease. This product is formulated for long-term storage and is sold at many large gun stores. Before storing metal, clean its surfaces—wear rubber or cloth gloves while cleaning and packing the device in grease so your sweat doesn't get onto the metal and promote rust.

Watertight containers stored in the ground have a tendency to create condensation moisture from the air trapped inside them. Placing a hygroscopic chemical packet inside the container away from the metal will overcome any rust problems such moisture might create. Silica gel is the hygroscopic chemical of choice. It's available at most drug stores or can be purchased from the Hydrosorbent Company (Box 437, Ashley Falls, MA 01222).

Ideally, the container you cache your valuables in will be buried with a thick layer of rocks or gravel under it. This will allow rain water to pool below the container and seep into the soil. Plastic sheets should be layered over the container to divert water away. The earth and plants over the cache should be carefully camouflaged.

If you're concerned about criminals or authorities utilizing metal detectors to find your cache, there's a simple countermeasure that will thwart their efforts somewhat. Just sprinkle bits of metal, aluminum foil, BBs, tacks, or scrap metal around the area. Although high-tech detectors can be adjusted to ignore such countermeasures, this will set off less-sophisticated detectors, making them ineffective. For a detailed look at how to counter government searchers using high-tech equipment, see Ragnar Benson's *Modern Weapons Caching* (available from Paladin Press).

As for the cache container, there's a wealth of products

The Low End

that can be utilized. One of the cheaper can be created by going to a hardware store and purchasing a short length of plastic sewer pipe. This can then be assembled with the proper couplings (also available at the store) to create a short tube with a screw-in plate on either end. The plastic can be cut to length with a regular handsaw and is easily joined and sealed with a solvent (available at the store). For a few dollars, you can create a cache tube that is watertight and big enough to hold a million bucks or a rifle that has recently been outlawed by your friendly local city commission.

Those wanting to do a little shopping in the surplus market can often locate waterproof containers that are suitable for caching. Just be sure the seals are still tight—some are for sale because they no longer are watertight!

Although not as inexpensive as the Homak security cabinets, used safes, or caches, several companies do offer low-end safes and heavier strongboxes that might be a consideration for some readers wanting to store their valuables inside their homes. One of these companies is the Tread Corporation (P.O. Box 13207, Roanoke, VA 24032), which offers gun storage cabinets with double keys as well as safes with combination locks. Prices vary from slightly over $200 for the 1900 series of gun cabinets to $1,000 for the company's 3600 series of safes.

Arkfield (P.O. Box 54, Norfolk, NE 68702-0054) also offers a full line of keyed security cabinets. Prices vary from their Model 1 with a $277 price tag on up to their $450 Model 4, which weighs 646 pounds and has an interior big enough to hide in (though this shouldn't be tried since you'd suffocate in a matter of minutes).

Viking Office Supplies (13809 S. Figueroa St., P.O. Box 61144, Los Angeles, CA 90061) offers a Meilink fireproof safe with an internal capacity of one cubic foot for $270 and a two-foot internal-capacity safe for $370. These weigh only 85 and 170 pounds respectively (light for safes), so some care needs to be exercised to anchor them so they

The Arkfield Model 1 (above) and Model 2 (facing page) security cabinets. (Photos courtesy of Arkfield Manufacturing and Distributing Company, Inc.)

can't be carted off. They have combination locks so you don't have to worry about carrying a key around, but they're not nearly as secure as a heavy safe—to be expected since they cost only a fraction of the price of a heavy safe. But they will protect papers and other valuables from a fire as well as many amateur burglars.

The Low End

Finally, one thing you might bear in mind is the fact that many criminals will be limited in what they can quickly cart off from your home. That means if you invest in a few "decoys" like junk TVs that look good but are actually worthless, an old safe painted to look new but is actually empty, fake jewelry, and so on, you may be able to trick a burglar into going to a lot of work lugging off some heavy junk instead of the valuables you've hidden inconspicuously elsewhere. Give it some thought.

The Arkfield Model 3 security cabinet (shown here) has an interior almost big enough to hide in. The Model 4 is even more massive. (Photo courtesy of Arkfield Manufacturing and Distributing Company, Inc.)

THE HIGH END

Those wanting maximum protection of their valuables at home will have to pay extra for it. Unlike security cabinets, bona fide safes can cost some money. But many are not as costly as one might think, and all merit consideration if you have expensive or sensitive materials in your home that must be protected from thieves, snoops, or other pests that are prevalent these days.

Most of the safes listed in this section use Sargent & Greenleaf combination locks, which have established a reputation for being hard for criminals to manipulate. Many of the safes listed in this section also utilize a "relocker" system that keeps the mechanism from opening even if attacked by a torch, sledge hammer, drill, or punch. This is an important plus and worth paying a little extra for.

Competition among safe manufacturers is fierce these days, with new features surfacing with great regularity. It's wise, therefore, to write to companies carrying safes you're interested in to see what they're currently offering. Also, some companies have a way of coming onto the scene, lasting a few years, and then vanishing. If you have any doubts about a company, go with one of the "tried and true" listed in this book.

Although most safes are exceedingly heavy, care still needs to be taken to anchor them down. Burglars with enough time and tools can cart away unbelievably heavy safes; once they do this, they have the time to attack the safe without fear of being apprehended. You lose not only the contents of the safe but the safe itself for an added headache.

One practice you should avoid is backing the combination off only slightly after closing it so you'll not have to go "through the numbers" to open it again. This practice is common among many safe owners—and criminals know it. The first thing they do when finding an unguarded safe is to try to manipulate the dial to see if the owner has been foolish enough to leave it in this condition. Don't defeat the high level of security you achieve with one of these safes by saving yourself a few seconds' time in opening it.

Also avoid writing the combination down *anywhere*, especially somewhere close to the safe. Again, standard operating procedure for criminals finding a safe is to look around for a scrawled set of numbers that will give them the "open sesame" to get to your valuables. It's also smart to avoid your street address, phone number, or other obvious choices when selecting a combination.

The most common safe these days is the "gun safe." This is actually just a tall safe modified internally with gun racks and carpeting to accommodate the valuable guns many collectors own. These safes make good buys for home use since they are readily available, don't attract much attention, and can be shipped almost anywhere in the United States.

Several of the manufacturers listed in this chapter also offer a safe door and frame that can be mounted in concrete to create a walk-in vault that rivals that of a small bank. For those wanting to have a huge storage area that is protected from fire and criminals, this is definitely the way to go. On-site construction of the reinforced walls, floor, and ceiling can easily be handled by most local construction companies, as can the installation of the door itself. Just be sure to follow the manufacturer's instructions so it's securely mounted and can't be pulled out!

The High End

Some of these doors can be mounted with an internal handle so the owner can enter the vault and close it behind himself. This would create the ultimate in "refuge rooms," provided some provisions were made to circulate fresh air into the vault. (It should be noted that even the doors without an internal release can be opened quickly by removing the back plate with a screwdriver and releasing the lock levers—a trick magicians have employed for decades to escape from safes dropped into rivers or the like.)

Now let's take a look at some of the major safe manufacturers of readily available gun safes.

One well-known company among shooters is Browning (Route One, Morgan, UT 84050). This corporation has added a number of products to its firearm line, including the excellent Pro-Steel series of safes. The top-of-the-line Deluxe Presentation safes vary in price from around $2,000 to $3,000; the company's Gold series costs from around $1,000 to $1,200; the Signature series from $705 to $3,000; and the Silver safes run from $615 to $1,150. Browning also offers the FR62 fire-resistant safe, which has a keyed combination lock and extra insulation that gives it an edge in fire protection.

The Browning Pro-Steel line boasts a massive locking bolt design with up to twelve internal bars controlled by the lock to hold the door in place even if the hinges are defeated. Prices vary according to size, internal shelving, finish, and the type of lock used. Also, options like 24-karat gold plating on the lock, key-locking dials, and so on, can run the price higher. Browning also offers a vault door and frame (80 x 30 inches) for those wanting to create an in-house vault.

Perhaps the best known of today's American safe manufacturers is Fort Knox (1051 North Industrial Park Rd., Orem, UT 84057). This company offers a wide range of safes, with the larger ones having the option of an inside release mechanism added to them (especially comforting

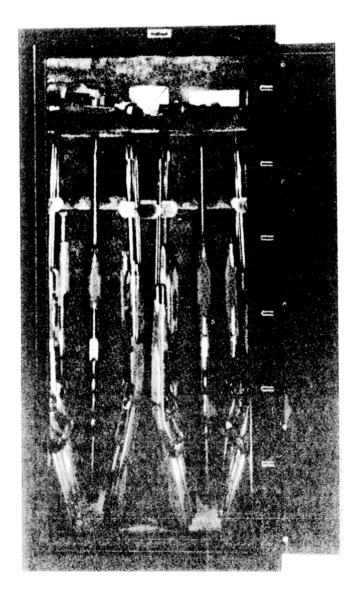

Two of Browning's Pro-Steel safes. Above is the 26-gun Sportsman; on the facing page is a 13-gun version of the safe. (Photos courtesy of Browning.)

The High End

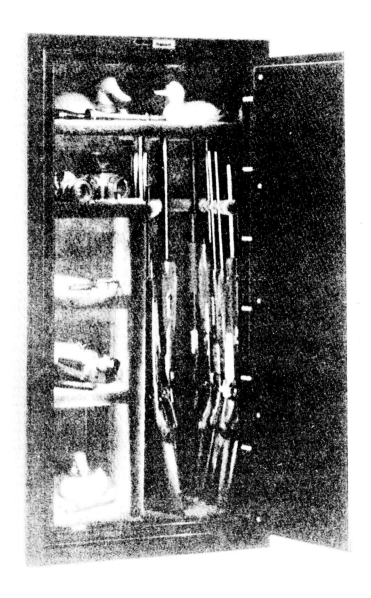

Like other safe companies, Fort Knox offers a wide range of models and custom features to create a safe for almost any need. (Photo courtesy of Fort Knox.)

for those fearful of being locked inside the safe). This same mechanism is also available for the company's vault door, making it ideal for those wishing to incorporate it into a refuge room or other shelter.

The top-of-the-line Fort Knox series is the Chuck Yeager Edition, whose namesake often appears in the ads for these safes. The Chuck Yeager safes retail from $4,520 to $6,000 and have locking bolts at the top, bottom, sides, and corners of the door for maximum resistance to prying. A multigear drive makes these units easy for the owner to unlock, and just about every feature a buyer might imagine is available for these vaults.

Fort Knox's Executive series of safes is similar to the Chuck Yeager but lack corner cross bolts and some of the other features of the more expensive units. Cost runs from around $3,000 to $4,000 according to size and options. The Guardian series, with price tags from $1,600 to slightly over $3,000, has fewer cross bolts and a thinner door. The Protector units run from around $1,300 to $2,800 according to size.

Also offered by Fort Knox are some relatively inexpensive fire safes. Though not overly resistant to attack by burglars, they do offer "fire insurance" for important documents. Costs run from $398 for a closet model (the 1050) to $78 for a locking strongbox (the 1010) that could be utilized inside a regular gun safe to protect documents from fires.

Liberty Safes (P.O. Box 50555, Provo, UT 84605) is another well-known company offering safes. The top-of-the-line for this company is the Presidential series of safes featuring locking bolts on the top, bottom, and sides of the door. Prices vary from $3,250 to nearly $5,000 according to size.

Other models offered by Liberty Safes include the Washington, Lincoln, and Franklin lines. Prices for the massive Washington safes vary from $2,600 to $3,700 according to the size and options. Prices for the Lincoln range from $1,800 to $3,000, and the Franklin, $1,060 to

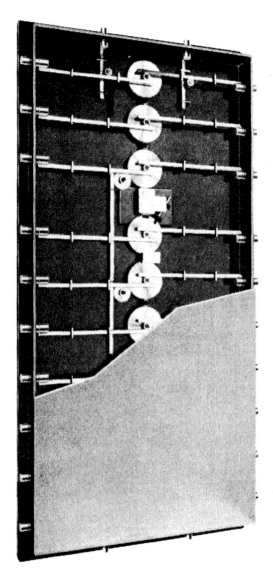

A cutaway photo of the internal workings of the Chuck Yeager safe door from Fort Knox. Note the complicated gear work that controls the locking bolts all the way around the door. (Photo courtesy of Fort Knox.)

The High End

Top-of-the-line Fort Knox safes feature locking bars on the top, bottom, sides, and corners, making them extremely "pry" proof. (Photo courtesy of Fort Knox.)

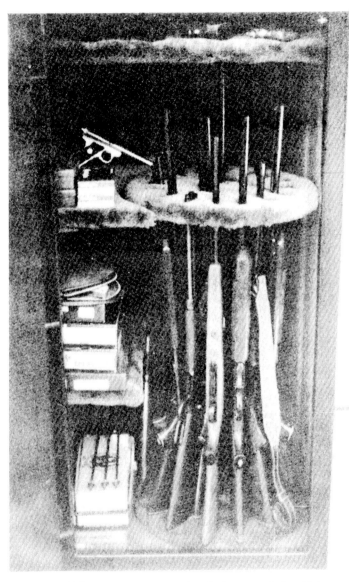

One of the many "custom" Smith Security Safes. This one features a lazy-Susan-style carousel that allows easy access to the rifles and shotguns stored in it. (Photo courtesy of Smith Security Safes.)

$1,800. Liberty offers extra fire protection in its Washington and Presidential units in addition to a larger number of locking bolts on the doors of these two models.

Smith Security Safes, Inc. (P.O. Box 185, Tontogany, OH 43565-0185) offers what amount to custom safes, generally constructed to the buyer's specifications (though it does offer a line of "standard" safes with a wide range of options). The company carries safes large enough to accommodate from sixteen on up to ninety-nine guns, with prices running from $470 to $1,350. Options such as paint or pinstriping for the exterior, interior shelving, fireboard lining, thicker steel plate, and chromed locks will quickly run the price up.

Smith Security Safes offers several interesting options. One is to trade in old safes of any manufacture to obtain a discount on one of the company's new safes. Another is a $5 "changing key" that allows owners to recalibrate the combination lock on their vault to a new series of numbers. This would be useful to those fearful of having the number compromised by an ex-employee who had access to the safe, a disgruntled spouse, or some such thing.

The Poor Man's Fort Knox

Like several other safe companies, Smith Security Safes offers doors and frames that can be utilized to create massive walk-in vaults that rival those of a small bank in both size and security.

CONCLUSION

Whether you're worried about a burglar stealing what you've worked hard to obtain or an overzealous government agent snooping into your affairs, a home safe can be the solution to regaining the security you desire. As we've seen, a good safe and a few simple precautions can give you a large amount of security for your treasures for a relatively low price.

But knowing how to protect your possessions won't guard them. Now you need to sit down and figure out what level of protection *you* need and can afford.

And then you must make it happen.

Good luck!